MINES of the EAST FORK

and its resorts

by

JOHN W. ROBINSON

LA SIESTA PRESS

1980

related La Siesta publications =

L Burr Belden
MINES OF DEATH VALLEY

Mary DeDecker
MINES OF THE EASTERN SIERRA

Helen Ellsberg
MINES OF JULIAN

Ronald Dean Miller
MINES OF THE HIGH DESERT

Ron and Peggy Miller
MINES OF THE MOJAVE

John W. Robinson
MINES OF THE SAN GABRIELS
MINES OF THE SAN BERNARDINOS

Pat Adler
MINERAL KING GUIDE

Ardis Manly Walker
KERN RIVER VALLEY VIGNETTES

ISBN 910856-69-9

LA SIESTA PRESS
BOX 406
GLENDALE, CALIFORNIA 91209

Contents

cover: Sedley Peck at Follows Camp
Sedley Peck Collection

SEDLEY PECK
Sedley Peck Collection

In Memory of Sedley Peck

'The Mayor of San Gabriel Canyon', so he was sometimes called. And with good reason. During his many years of living at Follows Camp alongside the sparkling East Fork, Sedley Peck was probably better known and more widely loved than any other canyon resident. He knew more of the canyon's history than anyone of his era. When he died, he left a good-sized trunk full of notes, mementos, newspaper clippings and photographs covering the rich saga of San Gabriel Canyon from the Indian days to modern times.

Sedley Clarendale Peck was far more than just a canyon historian. He lived a full life of adventure as world traveler, pilot, soldier in both World Wars, and newspaper correspondent.

Sedley was born in Los Angeles in 1889. As a boy, he spent many happy weeks at Follows Camp, founded and operated by his uncle and aunt, Ralph and Jennie Follows. From his grandfather, William T. Heaton, pioneer canyon gold miner, he learned the rudiments of prospecting. He was personally acquainted with many of the legendary characters in canyon history, including Oliver Justice, George Trogden, Henry Roberts, William Ferguson and Judge Henry O'Melveny.

Sedley attended Stanford University and earned two degrees in journalism. He was teaching high school in San Mateo when World War I broke out. Yearning for adventure and indignant with the Germans, he journeyed to France, learned to fly and became a combat aviator. Shot down by the Germans, he was briefly held prisoner but released because of his American citizenship. When America entered the conflict in 1917, Sedley transfered to the U. S. Army Air Corps and flew a Spad fighter plane on many missions over the Western Front.

Following the Armistice, Sedley joined the Paris office of the New York Herald. He became the newspaper's leading European correspondent and covered many of the leading stories of the era between the wars.

History repeated itself in World War II, as far as Sedley was concerned. He rejoined the French Army as an ambulance driver, was captured by the Nazis when France was overrun, but then was released after a short captivity. He again transferred to the U. S. Army when this country entered the war. He served as an intelligence officer on General Eisenhower's staff and rose to the rank of colonel. While stationed in North Africa in 1943, he met the charming young lady who was to become his wife.

With the war over, Sedley returned to the United States with his young French bride, Dolores, better known as 'Flo Flo'. They fashioned a home at Follows Camp on the East Fork in 1945. Here Sedley happily spent the last two decades of his life, active in such organizations as the American Legion, the Masons, as well as E Clampus Vitus. Together they entertained a steady flow of prominent guests, and collected canyon history. Kenyon De Vore, a close friend of Sedley and Flo Flo, recalls "Sedley was very high in the American Legion and active in Masonic affairs. During his East Fork residence, he and his Flo Flo were always going to some convention, where Sedley was often the speaker. He was a good writer and an excellent, fascinating public speaker often sought by various groups. The procession of friends and guests at their Follows Camp home reads like 'Who's Who' ".

Sedley frequently gave talks on San Gabriel Canyon History to various historical organizations and was planning to write his own history of the canyon when he died.

To the memory of Sedley Peck, and to his Flo Flo, gracious hostess of Follows Camp today, this book is affectionately dedicated.

Introduction

The three-pronged canyon of the San Gabriel River cuts a deep crease in the southern slope of the San Gabriel Mountains. Its watershed extends across the heart of the range, from Red Box Saddle, where the West Fork is born, eastward some thirty-five miles to the slopes of Old Baldy, genesis of the rugged East Fork. Halfway between runs the North Fork, flowing down from Crystal Lake Bowl and the slopes of Mounts Hawkins and Islip. Some twenty percent of the total precipitation that falls in the San Gabriles flows down this watershed.

Of the three forks of the San Gabriel Canyon, the East Fork is the deepest, the most precipitous — and the most storied. Here eminate a host of stories and legends rich in human drama. Indians, a century of gold mining, fishing yarns, unsolved murders, popular resorts, destructive floods and fires have all, at one time or another, played center stage.

A complete history of the East Fork would run into hundreds of pages. Books could be written on the mining saga, the resort camps, the colorful canyon pioneers. Presented here is a capsule history of the East Fork, an outline of what man has accomplished here.

A history such as this requires the assistance of many people. The writer is particularly indebted to Mrs. Sedley Peck of Follows Camp, who gave me full access to Sedley's trunkfull of canyon lore, and to Kenyon De Vore of Arcadia, whose great knowledge of the canyon is derived from living most of his life within its defiles. Also providing generous assistance were the friendly staffs of the Henry E. Huntington Library, the Azusa and Glendora public libraries, and to the U. S. Forest Service, Glendora office. This story could not have been told — at least not in nearly as much detail — without their kind help.

John W. Robinson
Fountain Valley, California
April 20, 1980

FLO FLO PECK AT FOLLOWS CAMP STORE, BUILT 1861
Sedley Peck Collection

Indians, Spaniards and Pioneers

On a lonely knoll that arises north of Azusa, just outside the mouth of San Gabriel Canyon, was the Indian village of Asuksangna (the present city of Azusa takes its name from this aboriginal village). For untold centuries before the arrival of the Spaniards, Indian peoples of Shoshoean linguistic stock had lived in Asuksangna and other rancherias throughout the San Gabriel Valley. Anthropologists have given the name 'Gabrielino' to these early people who lived south of the San Gabriel Mountains because of their later assimilation into the San Gabriel Mission community.

The people of Asuksangna and other nearby villages left their brush huts and paid frequent visits to the chaparral-clad mountains and canyons to the north. These mountains provided the Indians with food and materials for building and hunting. For food, they hunted deer and rabbit, and gathered acorns, seeds, edible roots, wild berries and nuts. They fished the mountain streams and dried their catches for use during the winter months. Chaparral was an abundant source of many necessities. Manzanita berries were pressed for cider and the leaves were smoked. Greasewood provided arrow shafts for hunting, while yucca fibers were utilized for nets and cords. Of particular value to the Indians was the California Coffee Berry *(RHAMNUS CALIFORNICA)*, once abundant in San Gabriel Canyon. They used this shrub as a laxative and a corrective to the astringent effects of their acorn diet, and as cure for rheumatism and other disorders.

Because floods have periodically swept the San Gabriel River, its course is not rich in artifacts. No remnant of any aboriginal village has yet been uncovered in the canyon. However, there is abundant evidence that Indians visited both the West and East forks of San Gabriel Canyon. Jim Roberts, son of canyon pioneer Henry Roberts was born in the canyon in 1868, had vivid boyhood memories of Indians camping at the Camp Rincon site as late as the mid-1870's. Camp Rincon was supposedly a favorite aboriginal resting place. According to Sedley Peck, Graveyard Canyon, a tributary of the lower

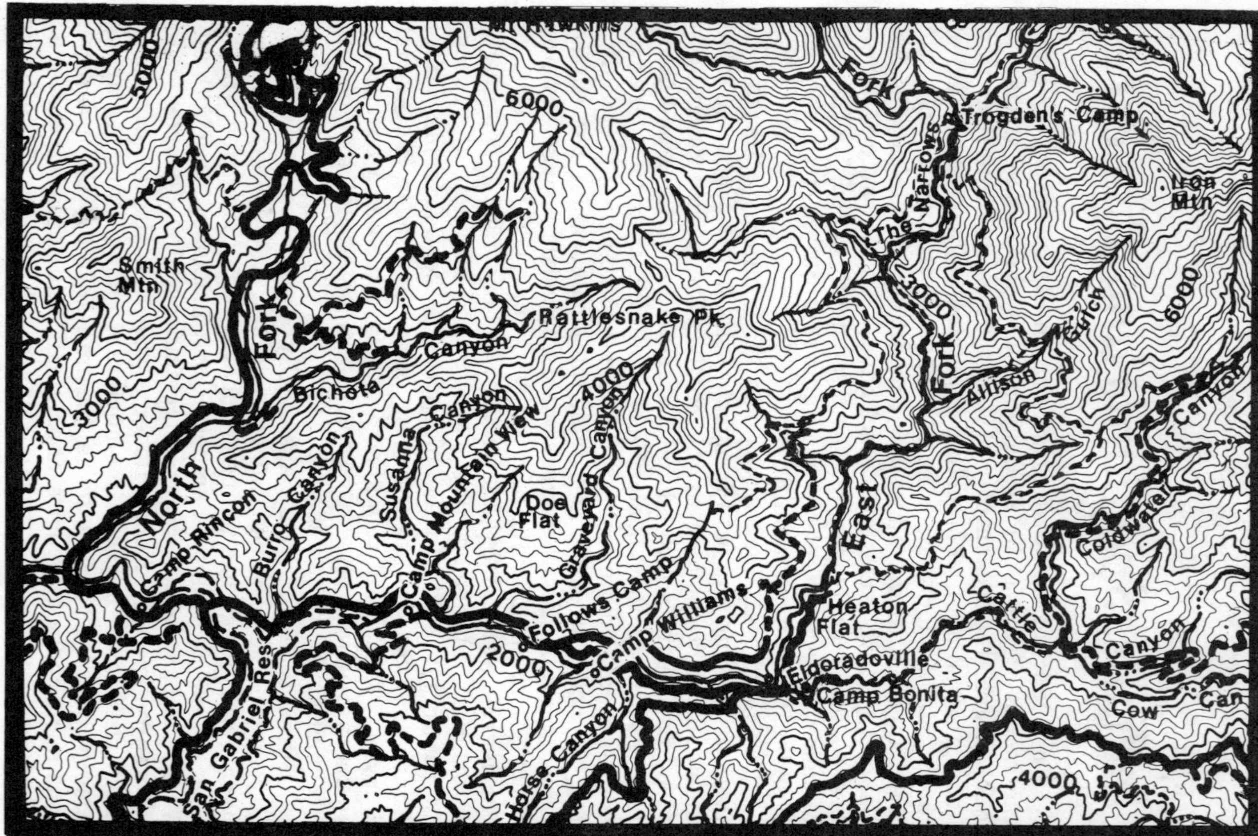
Fork
Trogden's Camp
Iron Mtn
The Narrows
3000
Fork
Allison
Gulch
6000
6000
5000
Smith Mtn
Fork
Rattlesnake Pk
Canyon
Bichota
Canyon
4000
3000
North
Camp Rincon
Burro
Canyon
Susanna
Mountain View
Camp
Doe Flat
Graveyard Canyon
Follows Camp
Camp Williams
East
Heaton Flat
Eldoradoville
Camp Bonita
Coldwater
Canyon
Cattle
Canyon
Cow
Can
2000
San Gabriel Res
Horse Canyon
4000

East Fork, was so named because of Indian burial mounds once located there. And some prospectors digging for gold near Camp Williams in 1938 uncovered what was believed to have been another ancient burial plot. Small artifacts have been picked up at the site of Follows Camp. These bits of evidence suggest that there may have been a Gabrielino village somewhere in the lower East Fork, Farther up the river, near the Iron Fork Camp above the Narrows, Indian artifacts were once abundant, according to Peck.

The Gabrielinos have been described as a "simple and happy aboriginal people who were well adjusted to the Southern California enviroment." Sadly, they could not adjust to the coming of the white man and his aggressive ways. The peaceful life of these Indians drew to a swift conclusion and, like all primitive cultures, they suffered greatly from civilization's encroachment. Extinction was the inevitable result. Today, there are no known descendents of the Gabrielinos who once frequented San Gabriel Canyon.

The end of the Indian way of life was heralded by the arrival of the Spaniards in 1769. Mission San Gabriel Arcangel was founded along the grassy banks of El Río Hondo in 1771. The Gabrielino Indians were incorporated into the mission community, and within a few years all but a handful of these once-content people were serving the padres, many of them by farming mission fields and tending mission herds. At the height of the mission period, during the first decade and a half of the 19th century, great herds of mission cattle, horses and sheep grazed the verdant grasslands of the San Gabriel Valley and slacked their thirst with the crystal waters of the San Gabriel River. Livestock must have often strayed into San Gabriel Canyon, to be periodically retrieved by mission vaqueros and Indian neophytes. It is probable that the mission herders were well acquainted with the canyon, both for retrieving livestock and for obtaining timber, at least as far up as the lower East Fork.

The mission era drew to a rapid close with the enactment by the Mexican government of the Secularization Act of 1833. No longer were the Franciscan padres masters of the mission lands or the Indians who toiled on them. The end of the mission era ushered in the golden age of the great private ranchos, the time of 'The Cattle on a Thousand Hills.'

The mission lands below San Gabriel Canyon became the great Rancho Azusa, granted to Luis Arenas in 1842, purchased by Henry Dalton two years later.

BILLY HEATON AND RALPH FOLLOWS, ca.1917
Sedley Peck Collection

Henry Dalton, born in England in 1803, set sail for South America at the age of fourteen. He spent the next twenty-five years of his life in Peru, amassing a modest fortune as a merchant. Dalton came to Southern California in 1843 and promptly bought several lots in the small pueblo of Los Angeles. But his real ambition was to become a ranchero and he was not long in satisfying this desire. On December 24, 1844 he paid Arenas $7000 and became owner of 4400-acre Rancho Azusa. Within three years he had purchased several adjoining ranchos, and his holdings extended along the foothills of the San Gabriel Mountains from present-day Pasadena to San Dimas. To complete his acceptance into the Mexican-Californio community, he married into the prominent Zamorano family, embraced Roman Catholicism and was rechristened Enrique Dalton.

Don Enrique Dalton built his hacienda, midway in his vast domain, on the Rancho Azusa de Dalton. Here, on a slight rise southeast of the mouth of the San Gabriel Canyon, he fashioned a thick-walled adobe home. The beams, rafters and shakes for the roof were hewn and split from the oaks and pines of the canyon.

Dalton was well acquainted with 'Azusa Cañon', as San Gabriel Canyon was then known. His cattle and horses grazed within its confines, as far north as the junction of the East and West Forks. He took timber from the canyon, and there are stories that some of his timber cutters recovered gold from the East Fork gravels. He used the canyon waters to irrigate his fields. In 1845 Dalton petitioned the Mexican authorities for title to the canyon and mountains lying north of Azusa. A committee of the Los Angeles ayuntamiento (town council) reportedly favored his request for "the mountains and cañons of the Azusa stream and Boca Negra, excepting San Dimas," but the grant was never made.

One of Dalton's ranch hands, Domingo Arviso, a Mexican vaquero, was the first white man known to have settled on the East Fork. In 1850, according to Sedley Peck, Arviso brought his bride directly from Arizpe, Sonora and settled on a shaded bench two miles up the East Fork, the spot known today as Mountain View. Here the Arvisos built a stout cabin, tilled the canyon floor, hunted, fished, and raised a large family. They were a familiar sight to canyon travelers until well into the 1870's.

The most noteworthy name associated with San Gabriel Canyon and the East Fork during the latter half of the 19th century was Henry C. Roberts. Roberts crossed the plains in 1849 to partake in the California gold rush. Although it was the lure of gold that initially brought him here, he was an opportunist and changed trades as the occasion warranted. He was, in turn, a miner, a merchant, a stage coachman, a real estate promoter and a land lawyer. After a short stint as a storekeeper in San Jose, Roberts moved to Los Angeles in 1857 and opened a livery stable.

It was stories of gold that first attracted Henry Roberts to San Gabriel Canyon. In the summer of 1859, Roberts and a partner named Williams inaugurated a stage line that made bi-weekly trips from Los Angeles to the East Fork gold placers (see following chapter). Later that same year, he built a house of rough board siding and shingled roof just outside the canyon mouth. Here he put in stock corrals and maintenance equipment for his stages and pack trains which supplied East Fork miners and the canyon residents for many years. In 1861 he opened a general store on the verdant streamside bench that later became Follows Camp. (This building, still standing behind Mrs. Sedley Peck's home at Follows Camp, is the oldest in San Gabriel Canyon.) The Roberts Store supplied the canyon miners with both necessities and luxuries. The medium of exchange was gold from the nearby diggings. In the early 1870's, Roberts invested a small fortune in hydraulic mining projects in the East Fork. He lost money on this enterprise, but recouped his small fortune with some fortunate Mexican mining investments. For some 55 years, until his death in 1914, Henry Roberts was a leading figure in and around San Gabriel Canyon. He was survived by his wife Susanna, who continued to live at the Roberts' canyon-mouth home until her passing at the golden age of 94 in 1934. The Roberts produced eleven children, several of whom, most notably Jim Roberts, were long active in canyon affairs. Roberts Canyon, just west of the mouth of San Gabriel Canyon, commemorates Henry Roberts today.

William Potter, carpenter and blacksmith, came into the East Fork in 1860 to try his hand at gold mining. He remained there the rest of his life. In the 1870's Potter acquired the Mountain View property from Arviso and built there in 1878, using hand-sawn lumber and split shakes, one of the most pretentious homes in the canyon. He was known to some of his friends as 'Lying Bill' for the tall tales he enjoyed telling. Potter mined, hunted, fished and cultivated his orchard and garden until his death in 1903.

The last two decades of the 19th century saw many others establish homes on the East Fork. There was big John MacCaslin, who fashioned a crude cabin near the mouth of the East Fork and ran a pack train up the canyon for many years. The brothers from Scotland, Archie and Duncan McNabb, the latter known as 'Stuttering McNabb', the stammering Scot, were familiar figures in the canyon for twenty years. Old Joe Everhardy perched his little cabin on a cliff above the mouth of Burro Canyon* and spent his time digging for gold and tending his goats. Fred Maley prospected and hunted for many years close to the great bend of the East Fork and is today remembered in the placename Maley Flat. Heaton Flat honors the memory of pioneer miner William T. Heaton (more on him later). For some reason the East Fork appealed to Yaqui Indians from Sonora, Mexico, four of whom settled in the canyon in the 1880's — José Maria, Don Philipe, Troubadour Rodriquez and one known only as Luisiena. It was all of these hardy individuals, along with others whose names have faded in the dim light of time, who tamed the canyon wilderness.

*Early Angeles Forest maps show it as 'Burrow Canyon. The derivation is unknown

CAMP RINCON STAGE
Sedley Peck Collection

The Gold of Eldoradoville

The saga of the East Fork's early days can be summed up in one word – GOLD. For more than a century, the precious yellow metal has been panned, sluiced, long tommed, hydraulicked and blasted out in quantities that have made San Gabriel Canyon one of the major gold producers in Southern California. Estimates of the total yield from the canyon vary from $4 million to over $13 million.

Just when gold prospecting began on the East Fork is difficult to establish. There are stories that Indians, working for their Spanish masters, dug for gold along the river long before the Anglos arrived, but these tales cannot be verified and must be viewed as fanciful myths.

The traditional story goes that some prospectors led by a Captain Hannager, drifting down from the Kern River mines, entered the mountains via Cajon Pass and penetrated to the headwaters of the East Fork. Here they found gold and stayed to recover small amounts. First news of the strike appeared in the *LOS ANGELES STAR* of September 21, 1854: "There has been some excitement this past week about the new gold diggings on the headwaters of the San Gabriel. We have met several persons who have been out prospecting and although they found gold of the best quality, differ very much as regards the richness of the mines. The Crab Hollow diggings are considered the best and will pay from two to five cents to the pan. It is understood that if the river can be turned from its present bed, some rich leads may be found and big piles realized."

For five years, minor placer mining activities continued on the East Fork, with a reported "average yield of six and seven dollars per day, to the man." Then, early in 1859, some prospectors struck it rich and the rush was on. By May, the East Fork was being prospected over almost its entire length, and promising new discoveries were made.

The *LOS ANGELES STAR* had sent a correspondent into the canyon to report on the excitement. On May 28 he wrote, "The cañon through which the San Gabriel flows has been prospected about forty miles up, and in every instance, the 'color, has been obtained . . . All who are in operation are taking out from $2 to $10 per dav to the man." By July 23, the *STAR's* correspondent stated that there were 300 men working in the canyon, including a large number of Mexican miners. And many continued to pour in.

In July stage service opened to the canyon diggings. The Roberts and Williams Stage Line offered tri-weekly trips, leaving Los Angeles at 7am and arriving at the mines by 4pm. The fare was set at $3 to the canyon mouth, and $6 to 'Prospect Bar' in the lower East Fork.

With the large number of men working in the canyon, it was not long until a mining camp took shape to supply the miner's needs and relieve them of their gold. The *STAR* of May 28, 1859 gave first mention of a rudimentary settlement consisting of a "boarding house, 2 or 3 stores, blacksmith shop, butcher shop, etc." The settlement seems to have been initially called Prospect Bar. Later it became known as Eldoradoville, and it was located at the elbow of the East Fork, near the latter's confluence with Cattle Canyon.

Mining continued at a hectic pace all summer and fall. A number of companies were formed to work claims along the river and on the slopes. These mining companies were usually organized by one man, who either paid wages or agreed to split the profits with the five to twenty miners working for him. Taking advantage of technical know-how and manpower, the companies began to use more profitable methods to recover gold. The *SOUTHERN VINEYARD* (August 23, 1859) described some of the methods used:

> Among the most extensive mining operations upon the river are those of the Little Falls Co. and McClure & Co. The former have constructed a flume several hundred feet in length, and of sufficient capacity to carry the waters of the San Gabriel at a high stage. All the modern facilities have been brought into requisition, viz waterwheels, pumps, derricks, railroads, etc., all of which they have accomplished by their industry in the short space of three months . . . The latter of the parties named above are engaged in putting up a hydraulic pump for the purpose of washing a hill claim.

A name mentioned often as one of the leading miners in the canyon was Thomas Driver. The *STAR* (Novenber 5, 1859) reported that "The Driver claim is being worked on an extensive scale. A dam has been constructed which lays bare a large section of the river bed, which they have found quite rich . . . This company clears $1000 for their past week's work."

Those who know the San Gabriel River are aware of its occasional intemperate moods, when floodwaters ravage the canyon bottoms. One very rainy night in late November 1859, the miners were introduced to this vicious aspect of the river. The *STAR* of December 3 describes the results: "We regret to have to record the total demolition of the mining works in the San Gabriel Cañon . . . So tremendous was the force of the torrent rushing down, that it swept away as chaff all the mining works erected on the river — dams, wheels, sluices, everything, in fact. The amount of damage substained by the miners cannot be calculated."

But the obituary was premature. Within the month, the miners were back at work, rebuilding their dams, waterwheels and sluices, and taking out as much gold as ever.

The *STAR* of March 10, 1860 reported that, "'There are at present eight companies engaged in bringing water by means of ditching and fluming of their claims on the different hills, so as to ground sluice or hydraulic wash, when practicable." The newspaper also announced that, as a result of a meeting of the miners at Falvey and Cullin's store, the Eldoradoville Mining District had been formed — the first use of the name 'Eldoradoville'. The mining district was defined as "Commencing at the junction of the waters of the main San Gabriel River and Cattle Cañon Creek, thence following the course of the main river toward the head, two miles above the claim known as the Nevada claim, and running on both sides of the river between said points and at right angles to the river course, to the centre of the main ridge on both sides," An elaborate set of mining laws, embracing 27 items, was enacted. The *STAR* correspondent concluded his report by predicting that "the returns of gold dust from the San Gabriel will be second to those of no other river in the state."

The town of Eldoradoville sprang to life after the 1859 flood, a rowdy successor to Prospect Bar. Sedley Peck described it as "The Downieville of the South — a rough, tough miner's town." The law didn't exist here; 'justice' was administrated by the muzzle of a six-gun or the blade of a knife. The *STAR's* correspondent complained, "We have frequent disturbances of the peace here and since we have no local officers, rowdyism and sanguinary assaults are of very frequent occur-

WIDMAN CABIN, COLDWATER CANYON
Robinson Collection

ance. There is no punishment at hand, parties get more bold in their nefarious practices. If death is not the result, there is no notice taken of the number of assaults with knife or pistol. At one o'clock yesterday morning, one Mexican or Indian killed another, by stabbing him in the breast with a knife. The apathy with which white men received this news was, to say the least, degrading to our sense of civilized refinement. But as a person is allowed to flourish with impunity deadly weapons, of course this must be expected."

Eldoradoville boasted three stores and, according to Peck, a half dozen saloons, with gambling and dance halls running wide open. John Robb, who allegedly spent more than sixty years in the canyon, told Peck that he "made more money by running the sawdust from the floor of the Union Saloon through his sluice box than he was able to make from real mining, so prodigal and careless of their pokes were the miners and gamblers of those days."

Surprisingly, Eldoradoville's miners showed great interest in politics. The crucial election of 1860 was watched with keen interest by the canyon prospectors. Most of them apparently were Democrats, with sympathies between the North and South closely divided. Democratic political rallies were held in the canyon during the summer and fall months, and several candidates for county and state offices felt it worthwhile to pay campaign visits to Eldoradoville. Miners and other canyon residents turned out in full force on election day. The results, published in the *STAR* of November 24, 1860, showed a clear victory for Democratic candidates. Among the presidential contenders, Northern Democrat Stephen Douglas led with 34 votes, followed by Southern Democrat John Beckenridge with 23 votes. Abraham Lincoln was a poor third with 14 votes.

During the Civil War years (1861-65) there apparently was some undercover pro-Confederate activity in the canyon. Gustov Brown, an Army detective assigned to watch secessionist movements in Southern California, reported that the Knights of the Golden Circle, a secret pro-Confederate organization, had a chapter in "The San Gabriel mines" with a membership of 27. Brown warned that the Knights had secret hideouts in the mountains, were well armed, and were prepared to open guerilla warfare when the appropriate occasion presented itself. No such 'occasion' ever took place.

Mining continued at a hectic pace throughout 1861 and into 1862. The *STAR* (August 17, 1861) stated that "Wells, Fargo & Co. reported the shipments of gold from their Los Angeles office during the last six months as averaging $15,000 a month," and most of it came from San Gabriel Canyon.

Eldoradoville died as it lived — violently. During the night of January 17-18, 1862, a torrential cloudburst hit the mountains. Early the next morning a flood of churning gray water swept down East Fork, obliterating everything in its path. As men and women of Eldoradoville scrambled up the hillsides to safety, the shanty town was literally washed away, lock, stock and barrel, as were the canyon bottom works belonging to the miners. Shacks, whiskey barrels, groceries, beds, roulette wheels, sluices, long toms, wing dams and China pumps were swept clean out of the mountains into the floodplains of the San Gabriel Valley. Where the day before there was a town, nothing remained but mud and boulders.

Thus ended the boom days on the East Fork. The miners headed elsewhere, most of them for the new strike along the lower Colorado River. Eldoradoville was never rebuilt. Although mining continued in the canyon until recent years, it never again approached the scale of the 1859-62 period.

The Great Hydraulic Race

The canyon's natural stillness is broken by sounds of feverish activity. A dozen men struggle with a lengthy canvas hose while two others aim a cumbersome iron nozzle — called a 'monitor' by the miners. Underneath, a group of Chinamen position the wooden sluice boxes. Finally, all is ready. The foreman shouts an order to the reservoir crew atop a hill, and within seconds a powerful spray of water surges from the monitor directly into the cliff. Boulders, gravel and dirt abruptly melt before the eyes. Immense chunks of earth collapse in watery heaps, then churn downward into the long sluice boxes. Below, the river turns a sullen brown.

After fifteen minutes the white surge lessens and gradually trickles to nothing. The hilltop reservoir is exhausted of its contents. As the artifical pond slowly replenishes, fed by a wooden flume that brings water from five miles up-river, the miners pick clean the riffles of the sluice boxes. Flecks of shiny yellow metal are carefully gathered and deposited in leather pouches.

Then the whole process starts over again, to be repeated a dozen times a day.

A similar drama is taking place just across the river. Crews on both sides of the canyon work twelve hours a day, now that water is high on the East fork. The year is 1872. Henry C. Roberts and William Ferguson are locked in their famous hydraulic mining race to recover the golden treasures of Southen California's most auriferous watercourse.

Henry Roberts came into San Gabriel Canyon in 1859, attracted by reports of gold there. In 1861 he built a store along the lower East Fork — later the site of Follows Camp. Located on a mesa just south of the river, Roberts store was one of the few buildings to survive the great torrent of 1862, and another flood almost as bad in 1868. Roberts watched the steady stream of nuggets which came to his scales from the Cecil Graham Hill claim immediately across the river, and finally decided to purchase the property and exploit it with hydraulic machinery. This occured in 1870.

EARLIEST KNOWN PHOTO OF SAN GABRIEL CANYON
ROBERTS' HYDRAULIC MINE, EAST FORK, ca.1873
Henry E Huntington Library Collection

A short time later William G. Ferguson, prosperous Los Angeles merchant and livery stable owner, purchased a hillside claim just below and across the river from Roberts' claim. Ferguson also had his eye on hydraulic mining. He and several others formed the San Gabriel Mining Company and commenced exploratory work.

Hydraulic gold mining — washing down auriferous hillsides with a powerful jet of water — was an established mode of operation in the Mother Lode country of northern California, but the process had received only fleeting use in the southern half of the state, mainly because of a chronic shortage of water. Only the largest watercourses provided sufficient flow to support major hydraulic enterprises, and then for only part of the year. Fortunately for Roberts and Ferguson, the East Fork was a plentiful stream for four or five months each year.

In order to 'hydraulic' effectively, the water must have a great force, or 'head' in miners' language. For Roberts and Ferguson, this required the construction of conduits from far up river and artificial reservoirs a thousand feet above the mines. The volume and pressure thus acquired was necessary to force water through the hoses and monitors with sufficient power to tear away the gold-bearing rocks, gravel, clay and dirt, then wash these ingredients through long sluiceboxes to recover the gold.

The walls of the East Fork are precipitous and unstable. No local surveyor or engineer could be found to undertake the task of plotting the course and grade for the conduits and supervising their construction over such difficult mountain terrain. Finally Roberts was able to acquire the talents of an English engineer named Sam Hawley. Ferguson found a surveyor by the name of W. W. Woodman. By the fall of 1871 the surveying was completed and the race was on.

The narrow canyon of the East Fork reverberated with the sounds of pick, shovel, hammer and blasting powder as both crews, working long shifts under the close supervision of Hawley and Woodman, hewed the ditches and built the flumes for the paralleling conduits. Roberts' flume commenced at Rattlesnake Gulch, just below the Narrows, and followed the west wall of the canyon high above the river, five miles in length. Ferguson's conduit left the river at Good Hope Bar and snaked its way along the east slope, bridging the mouth of Castle Canyon, four and a half miles in length.

No one knows just how it began, but a spirited rivalry erupted between the crews of Roberts and Ferguson. Some rather

FLO FLO AND SEDLEY PECK WITH HENRY ROBERTS' MONITOR
Sedley Peck Collection

substantial bets were made over who would first complete their conduits and commence hydraulic mining. The two claim owners were on none too friendly terms, allegedly because Ferguson had jumped a claim sought by Roberts and hired away some of the latter's Chinese miners.

Sedley Peck described the great race: "Dams were built to divert the river water into settling basins where the sand, which would have quickly clogged the ditches, was removed. Conduits were blasted and hewn from the rock cliffs of the canyon, flumes on steel supports driven into sheer walls, carried on high trestles across gulches and canyons, through clay lined ditches across high mesas, by gentle grades which soon raised the level, until the water was finally brought to the storage lakes a thousand feet above the rapidly falling canyon floor."

The hectic canyon activity caught the attention of the Los Angeles newspapers. The *STAR* sent a correspondent into the East Fork to report the race. On October 31, 1871 he wrote, "Times on the San Gabriel River are getting lively. Henry Roberts and Co. have finished four miles of ditch, mostly heavy work, used over 30,000 feet of lumber for fluming, got the hydraulic ready for action, and will commence piping in a day or two." Several weeks later (December 8), the correspondent reported, "San Gabriel Mining Company [Ferguson] finished a ditch four miles long, covering 200 acres . . . building a suspension bridge 200 feet high to carry the flume across Cape Horn Canyon . . . Roberts' flume is almost completed . . . A portion of the flume is suspended on the face of an overhanging cliff with iron bars drilled into solid rock.

Roberts was the first to complete his conduit, which was considered by Los Angeles historian J. M. Guinn to be "one of the most complete and substantial ever constructed in any mining camp in the state." Ferguson finished about two weeks later. Both conduits were completed in amazingly short time considering the difficult terrain, a tribute to the determination of the two mine owners, the hard work of over 100 Mexican and Chinese workers, the spirit of healthy competition, not to mention the estimated $60,000 expended on the dual efforts.

In January 1872 Roberts and Ferguson each turned on their hydraulic monitors, accompanied by the cheers of spectators assembled for the occasion, and commenced water-blasting away their hillside claims. Initially the ventures appeared to be highly successful as each recovered substantial quantities of gold — $5000 the first month, it was reported. Later they were averaging a more modest $1000 per month, according to Peck.

The apparent success of Roberts and Ferguson encouraged others to try their hand at hydraulic mining. The names of Matfield, Crow, Higbee, Ford, Caley and Justice all appeared in Los Angeles newspapers, all bent on duplicating the efforts of Roberts and Ferguson. But the most noteworthy of the later hydraulic ventures was the famous fiasco of 'Uncle' Dave Buell. Buell secured a hillside claim just down-river from Roberts, allegedly paying $40,000 for it, but could not gain water rights to the East Fork. So after some weeks he decided to convey water to his claim from the untapped North Fork. He hired a young engineer named Simpson to survey his ditch line. To get water from the North to the East Fork, Simpson proposed a 700-foot tunnel through the ridge separating the two watersheds. Buell agreed to their plan and work commenced on the tunnel and flume. Canyon pioneer Jim Roberts, son of Henry Roberts, described the results, "Great preparations were made for the opening of the works and visions of unlimited nuggets of gold crowded Buell's mind. Then the intake gates were opened and watchful waiting was the order of the day. But night fell and still no trace of water appeared. Buell betook himself up along the conduit and walked through the dry tunnel. A few hundred feet from the upper portal he heard the sound of rushing waters and discovered that Simpson had lost the grade." The tunnel had been bored uphill! Simpson left in disgrace and Buell allegedly went broke and departed for Mexico. The long abandoned tunnel, partly caved in and both portals hidden in brush, still lies high above the forks of the San Gabriel.

Despite repeated breakdowns in their ditches and flumes, Roberts and Ferguson continued to work their claims. Roberts apparently was the more successful; in the summer of 1873 he turned down an offer of $250,000 for his claim by a Chinese Tong from San Francisco, according to Peck.

Roberts should have accepted the offer. As it turned out, the returns in gold gradually decreased as the powerful monitors blasted deeper and deeper into the hillsides. And then, legal problems arose. The large scale hydraulic operations were sending down floods of muddy water into the populated San Gabriel Valley. discoloring domestic supplies and clogging irrigation pipes. Valley residents secured a court injunction against the hydraulic miners in the summer of 1874, and this soon was followed by state laws designed to prevent the pollution of domestic water supplies. Roberts, Ferguson and the rest of the canyon hydraulic gang were obliged to close down their gold-seeking enterprises, and all but minor, non-polluting hydraulic efforts ceased forever in the San Gabriel canyon.

GEORGE TROGDEN'S CAMP AT IRON FORKS, ca 1905
Sedley Peck Collection.

Gold is Ever the Lure

Although overshadowed by more prosperous mining ventures elsewhere, gold recovery along the East Fork continued almost uninterruped into recent years. With a few exceptions, after the end of hydraulic mining in 1874, these have been small-scale efforts by individual prospectors, many of whom built homes in the canyon.

Placer activities in the late 1870's and early 1880's apparently were sparse; newspapers of that period contain only fleeting mention of canyon mining. The Los Angeles *TIMES* of January 1, 1885 dismissed East Fork activity with the terse statement that, "For many years gold mining has been carried on in a small way in San Gabriel Cañon, but that lately has been abandoned to Chinamen and others who are satisfied with a small per diem return for their labor."

A name mentioned often during this period was Louis Sharp, who settled on the little bench halfway up the main canyon, known today as Sharps Flat. Sharp was given credit for discovering the silver vein high on Silver Mountain that became the Kelsey Mine, and later opened a silver and gold mine of his own just north of the Kelsey claim. The *TIMES* of June 29, 1882 reported excitement in town over a rich lump of pure silver from Sharp's mine on exhibit at the Pico House, Los Angeles' finest hosterly at the time.

In the late 1880's a flurry of new discoveries brought a small horde of miners into the East Fork. The *TIMES* of March 28, 1888 reported the incorporation of the San Gabriel Mining Company, capitalized at $1 million, to work the riverbed below the East Fork Narrows. J. S. Briggs was elected president, Henry C. Roberts vice president, and Joseph Cook superintendent of operations. Old Joe Cook was long a familiar figure among the East Fork diggings, in the canyon since the 1860's. He was a miner of visionary schemes. Sedley Peck described one of his ambitious projects: "Cook and several associates started a tunnel to Devils Canyon, the plan being to take the river underground and leave the workings free from water. These hardy miners worked in deep and swiftly running cold water, timbering with laurel logs as they went, for three heart-breaking years, but all to no avail since their tunnel developed more and more water as they went along, and their herculean efforts were doomed to failure." Following his tunnel fiasco, Cook devoted his efforts in behalf of the San Gabriel Mining Company. Eventually the company's claims covered two miles of the East Fork below the Narrows.

By 1889, the East Fork placers were "attracting considerable attention," according to the *AZUZA NEWS* (March 2, 1889). Prospectors were working almost the full length of the canyon. Henry Roberts, ever the enterprising businessman, placed the following advertisement in the *NEWS* (March 30, 1889):

> HO! FOR THE SAN GABRIEL CANYON!
> I am prepared to furnish burros and horses to all points on the river.
>
> Henry C. Roberts

Roberts' stable and pack train headquarters at the canyon mouth was deluged with business. His string of heavily-loaded mules was long a familiar sight to every canyon dweller.

By July 1889, there were 300 miners in the canyon, taking out an average of $2.50 to $3.00 in gold per day. By September, the average take was reported as $5 to $8 a day. Roberts was hauling a thousand board feet of lumber per month up the East Fork for sluicing. Two rustic mining settlements sprang to life: 'Camp Good Hope,' just below the East Fork elbow, and 'Summit City,' four miles above the narrows. Oliver Justice and Dr. V.K. Purdy were working the richest claims in the Narrows.

In 1893 the Upper San Gabriel Mining District was organized at Ralph Follow's store in the lower East Fork. The district encompassed all the placers claims on the upper river. William Potter of Mountain View was elected recorder. Later that same year, Jim Roberts was selected recorder of the new Lower San Gabriel Mining District, covering gold and silver claims from Azusa to the Forks.

Thanks to the memory and research of Sedley Peck, we have vivid descriptions of many of these colorful characters who mined the canyon during this period.

William Tecomseh Heaton, born in Virginia in 1837, came west during the California gold rush of 1849. After years of prospecting and drifting from place to place, he came into the East Fork in 1891 and, after prospecting for a while in Cattle

Canyon, settled on the mesa just north of the elbow known then as Peachtree Flat. Later it became known as Heaton Flat. He filed a claim there in 1902 and built a crude shack of stone and wood. He is said to have worked his mine "every day, rain or shine" until his death in 1924.

Oliver Justice, the lanky, long-bearded patriarch of the canyon, mined in and above the Narrows from 1870 until his death in 1924. His grave lies on a mesa above Iron Fork.

Alonzo Shoemaker, another bearded old-timer, mined the high bar above the small tributary of the East Fork that now bears his name. He was a familiar sight for twenty years, from 1870 to 1890. In 1890 a company of four miners leased the Shoemaker claim and installed a 'self-shooter' — a kind of miniature hydraulic system. The results of this unique experiment were disappointing and the effort soon abandoned.

Fred Manley came into the canyon in the early 1890's and worked a claim at the junction of the East Fork and Cattle Canyon. He is reported to have taken out $60,000 in gold before being served with an injunction for muddying the river waters.

There was Harry Walton, the 'Bee Man,' who lived for twenty-five years on the little flat across the river from Camp Rincon. Every summer he sold honey to canyon visitors; in winter he worked his gold mining, driving a tunnel deep into the mountainside in quest of the elusive metal.

George Trogden was described as a close likeness to Long John Silver, "even to his rolling gait, stiff leg and bad eye." Trogden entered the canyon in 1898 as an employee of the Pacific Light and Power Company. He enjoyed it so much that he stayed the rest of his days. For many years he worked a claim near the junction of Iron Fork and the East Fork, just above the Narrows. Here he built a stone house and grew his own fruit and vegetables, and was known throughout the canyon for his warm hospitality, "always ready with coffee pot and frying pan."

Every mining region had its fabled badman, and the East Fork is no exception. He was John Knox Portwood, a self-styled 'King of the Canyon,' prospector, claim jumper, pugilist, part-time packer, card shark, and, some said, a real lady-charmer. Portwood came into the canyon in 1895 and built a crude cabin in Cattle Canyon, just above its junction with the East Fork. He made his living by gold mining and occasionally running a pack train.

According to Sedley Peck, Portwood wanted to be known as a killer and boasted of the six notches carved in the handle of

DEPRESSION MINES IN SAN GABRIEL CANYON [Hooverville]. 1931
Henry E Huntington Library Collection

his six-shooter, supposedly denoting men he had shot in Virginia, Texas, Arizona and California.

On Christmas night, 1917 after a poker game at Trogden's Iron Fork cabin, Portwood allegedly shot and killed a young German miner named Herman Miller. Portwood was arrested by a sheriff's posse after Miller's bullet-ridden body was discovered in the bush a short distance from the cabin. Before his trial, the canyon badman offered a fantastic yarn that Miller was a German spy and that the Kaiser's secret service had an espionage base in the canyon! (World War I was raging, and anti-German feeling was at its peak in the U. S.) Portwood pictured himself as a patriotic citizen defending the cause of liberty! He told federal officers that Miller's cabin on the East Fork was often used as a rendevous by one Fritz Schlenberg, who was then under arrest in San Francisco as a German spy. According to the *AZURA POMOTROPIC* (January 4, 1918), the federal officers searched the Miller cabin and combed the East Fork, but found no evidence to back up Portwood's allegations. The spy story was not used at Portwood's trial. Instead he claimed self-defense. The badman was acquitted.

Portwood died as he had lived — by the gun. On June 12, 1920 two rangers, Charles Novatny and Jack Dunne, called on Portwood to investigate threats the latter had allegedly made against two canyon women. The badman suddenly drew his revolver and pointed it at Dunne. The rangers ducked for cover and a gun battle ensued. When it was over, 'The King of the Canyon' lay dead in the doorway of his Cattle Canyon cabin.

Another ornery canyon miner was old Joe Everhardy, who had a claim on the canyonside above the Narrows. In 1911, the Pacific Light and Power Company, planning a power project in the upper East Fork, built a trail around the Narrows that crossed part of Everhardy's claim. Old Joe, working his mine 200 feet above the new trail, became incensed at what he considered trespassing. To protect his privacy and his 'rights,' he developed the nefarious practice of rolling boulders down on passers-by. This dangerous game was brought to an abrupt halt by the sheriff, who acted on a complaint filed at Follows Camp. It was alleged that a horseback party from the resort camp was leisurely riding the trail when Everhardy began to hurl boulders down upon them, causing injury to a horse and narrowly missing several riders.

There were countless other colorful characters who, over

the years, sought a fortune in the East Fork. Among the many names are a few who stand out because of their originality — One eyed Charlie, Twitchlip Kelly, Two Gun Rosenkrantz, Charlie the Chink, Peg Leg Coynes and Soldier Thompson. What tales the ghosts of these departed miners could tell!

During the depression years of the 1930's, hundreds of unemployed men and their families moved into the East Fork to pan, wash and dig for gold. Two shanty settlements sprang up alongside of the canyon gravels. One, located at the site of Eldoradoville, became known as 'Hoovertown'. The other was at Mountain View on the lower East Fork. The residents there called it Mountain View; others knew it as 'Shantytown' or Murray Flats, in honor of the self-appointed 'mayor' of the group who had squatted there. Families fashioned crude homes from five-gallon tin cans, cardboards, tree branches and burlap bags. Many lived in canvas tents. Some resided in their automobiles, often with a canvas canopy attached to one side. These people, on the whole, were proud and independent. They would rather dig, pan or sluice gold from the river and its adjacent hillsides, recovering possibly $30 a month, than go on welfare. Some of their mining methods were quite ingenious, involving the use of pulleys, make-shift ore buckets and automatically-triggered sluiceboxes. Two determined individuals even hauled a mammoth steam shovel all the way up the East Fork to a point just below the Narrows. They used the cumbersome piece of equipment to pick up and dump tons of sand and gravel into a large sluice box alongside the river, thereby recovering appreciable amounts of gold through this unique effort.

The great deluge of March 1938 swept like a monstrous tidal wave through the canyon, completely obliterating Hooverville, a good part of Shantytown, and all the mining works. Fortunately few lives were lost, as there had been ample warning of the rising water. Thus abruptly ended the depression mining era. Only a handful of the depression miners returned to the East Fork after the flood.

In recent years, the East Fork has seen amateur weekend prospectors try their luck at goldpanning They do this mostly for fun, and if they happen to trap a speck of shining metal in the bottom of their pan, they celebrate with a can of beer. The mining in the East Fork today is predominately a recreational activity, although with recent increases in the price of gold, it is not beyond the realm of possibility that full-scale professional mining may yet return to the canyon.

Mountainside Mines

Not all of the East fork miners confined their efforts to the river and its bordering slopes. A few hardy prospectors scambled high up on the mountainsides, searching for the illusive 'mother lode' that supposedly fed the rich placers of the East Fork. While none ever found a bonanza, many did locate quarts claims and recovered varying amounts of gold. The Big Horn high above the East Fork headwaters, the Native Son of Prairie Fork, the Stanley-Miller above the East Fork Narrows, the Allison high in Allison Gulch, the Baldora in Dry Gulch, the Eagle and the Gold Dollar far above the head of Coldwater Canyon — these represent incredible feats when you consider the elevation, the rugged terrain, inaccessibility, and vulnerability to the elements. Most of these gold mines had a stamp or ball mill perched on a mountain shelf nearby, with heavy machinery for crushing ore. The task of lugging this equipment, piece by piece, up nearly vertical slopes and hauling out tons of crushed ore required a high degree of fortitude and strenuous work. Of stern stuff these mountaineering miners were made.

The 'granddaddy' of all these mountain mines was the Big Horn, perched at almost 7000 feet on the rugged east face of Mount Baden-Powell. Charles Tom Vincent, prospector and hunter who lived in a rustic cabin high in Vincent Gulch above the East Fork headwaters, discovered the gold-bearing quartz vein in 1895. He named the mine for the Bighorn sheep he had been hunting the day of his discovery.

Vincent did not have the money to develop the mine and it soon fell into the hands of promoters. It was not long until carefully-fostered rumors were heard to the effect that the Big Horn held fabulous riches within her rocky bosom. Stories circulated that the mine was "the richest in the world . . . Its vein is a mile wide . . . There is ore so rich that it can be sent to the mint without milling." Los Angeles businessmen were told that a $50,000 investment would return millions in gold. Bank President Jackson A. Graves related how a number of prominent citizens, including Gov. Henry T. Gage and

BIG HORN MINE, ca.1975
Robinson photo

U. S. Senator Stephen M. White, were on the verge of buying Big Horn stock when Graves, skeptical, urged caution. A mining engineer had told him that ore samples being shown about town showed evidence of 'salting.' To prove or disprove the fantastic claims of the Big Horn promoters, Graves and his engineer friend hiked up to the mine to check it for themselves. They removed ore samples from various locations within the mine and sent them down via Ralph Follows' pack train, using round-the-clock guards to prevent any possible tampering. An independent assay in Los Angeles bore out Graves' suspicions. The assay checked out at a miniscule $1.37 in gold per ton! The Big Horn's inflated bubble was punctured.

Nevertheless, there was gold deep inside Baden-Powell, if not in the quantities first proclaimed. During the years 1902 to 1910, the Lowell and California Company, a Los Angeles-based concern, invested a small fortune in the development of the mine. Tunnels and shafts were bored hundreds of feet into the mountainside. A 10-stamp mill was erected at the site, and a narrow wagon road was hacked across the cliffs to Vincent Saddle, then down to the desert. Gold-bearing quartz was blasted from the veins deep in the mountain, then loaded onto small ore cars and rolled out to the mill for crushing. Large horse-drawn wagons then transported the crushed ore out of the mountains for processing.

In the end, the Lowell and California Company met defeat. The $200,000 in gold recovered failed to cover the high costs of development, the expected mother lode could not be located, flooding deep within the mine caused insurmountable problems and the mill suffered a disastrous fire.

There was a brief renewal of activity in the 1930's, when The American Metal Company leased the Big Horn and put in a 50-ton flotation mill. Another mine flurry took place in 1941, but the war closure of gold mines ended it. Today, with the high price of gold, exploratory work is being done at the mine and there is the very real possibility that the Big Horn may again hum with activity.

The discovery and development of the Big Horn caused prospectors to search out other promising quartz veins high above the East Fork.

In 1897 the Native Son Mine was discovered and developed on the north slope of Prairie Fork, near the East Fork headwaters. Originally there were seven claims, but in 1900 they were consolidated into the Native Son Mining Company. Six tunnels were bored into the mountainside and a 3-stamp mill was erected nearby. Later, after the demise of the Victoria Silver Mine in the main San Gabriel Canyon, the 10-stamp Victoria mill was disassembled and dragged, piece by piece, all the way up the East Fork, through the Narrows, to the Native Son. Some of the pieces, made of cast iron, weighed several tons. It seems a marvel that such a task could be accomplished without benefit of wagon road or even a decent trail. The Native Son proved to be a disappointment and was shut down in 1920.

Perched precariously on the cliffs above the East Fork Narrows was the Stanley-Miller mine. Gordon Stanley and Ben Miller discovered promising quartz veins while scrambling along ledges in 1915, and promptly filed seven claims. Eventually 12 claims were patented, extending a half mile along the cliff. While Stanley worked the mine, Miller built the airy 'Wetwater Trail' that crossed the near-vertical wall above the Narrows. The trail was three feet wide and relatively safe — as long as one didn't look down to the floor of the East Fork, 300 feet straight down! But at one spot, where water seeped from the rock face and dripped onto the pathway, was the terror of the pack mules. Loaded with supplies for the mine, they often balked and tried to turn around rather than to traverse under the seeping water, much to the disgust of Miller. However, via this trail, equipment was packed in to build a cabin and a small ball mill outside the mine, clinging to the rock face.

Miller sold out to Arthur Fullwood in 1930 but Stanley remained, doggedly working the mine. Between 1933 and 1939, with the help of Fullwood, he managed to take out several hundred tons of ore. A 1939 accident sent Stanley to the hospital; he never returned to the mine. The cliff-hanging Stanley-Miller was abandoned to the elements. The cabin and the mill building were destroyed in the September 1953 forest fire that ravaged the upper East Fork. A Forest Service crew was dispatched to the Narrows to fight the conflagration and spent the night at Iron Fork Camp. During the night, flames reached the mine and exploded many abandoned blasting caps, creating an awful din. Suddenly the mill machinery wrenched loose and rolled all the way down to the bottom of the Narrows with a terrifying roar, routing the forestry crew from their bags. The Stanley-Miller met its end in spectacular fashion.

In 1913 John James Allison, veteran desert prospector, scrambled up the steep slopes above Laurel Gulch and then "crawling on his hands and knees through the brush," discovered gold-bearing quartz at the 5200-foot level. Allison and his three sons expended almost superhuman efforts to develop the mine high up on Iron Mountain. They managed to drag heavy cast-iron parts for a ball mill, rails and ore cars up the mountainside, utilizing an ingenious system of cables and pulleys to hoist the equipment the final 300 feet to the cliff-perched mine. They built several small cabins nearby. To get the ore out, they laborously hacked out a contour trail around the mountainside and down into Coldwater Canyon. The Allisons turned down an offer of a million dollars for the mine in 1915. They should have accepted. The ore veins pinched out and the Allisons never recovered their original investment. The family — particularly son Arthur Allison — continued to work their claim through the 1920's with little to show for their strenuous efforts. Around 1930 they finally gave up and a succession of leasees and claim jumpers tried their luck until 1942. According to the California Bureau of Mines and Geology, 10,000 tons of ore were recovered from the Allison Mine, yielding nearly $50,000 in gold.

Far above the head of Coldwater Canyon, just under the jagged ridge that connects Iron Mountain with Mount Baldy, were the loftiest and most inaccessible gold mines of all — the Gold Dollar and the Eagle. These alpine gold prospects were discovered in 1909 and 1913 respectively by Charlie Smith, long-time San Gabriel Canyon resident and founder of Camp Rincon. The Gold Dollar, perched near the ridgetop, had a spectacular aerial tramway that carried ore from mine to mill several hundred feet below. Both mines had small ball mills spotted on small ledges and held fast by steel cables. How this heavy machinery was hoisted so very far up the mountainside defies imagination (much of the machinery is there today, a source of astonishment to the occasional hiker).

Charlie Smith met his end in spectacular fashion. In January 1933, during a snowstorm, an avalanche swished down on his cabin at the Eagle Mine, carrying the sleeping Charlie, his bed and his cabin down the mountainside into Coldwater Canyon. He was buried in a tomb of frozen white. His remains were dug out a few days later by a search party.

After Smith's demise, Ed Geyer worked the mines for several years. Despite herculean efforts, neither the Gold Dollar nor the Eagle were more than marginal successes. They were abandoned just before World War II.

A thousand feet lower than the Gold Dollar and the Eagle, in Dry Gulch above Coldwater Canyon was the Baldora Mine, also discovered by Charlie Smith. After Charlie's death, the Baldora was taken over by Jake Widman and renamed the Widco. The mine never returned a profit, as far as is known.

Although none of these mountainside-clinging mines ever yielded a bonanza, one cannot help but marvel at the fortitude of these hardy souls who attempted what seemed like the impossible.

Mecca for Vacationers

"In the old days before hydroelectric development and flood-control dams, the San Gabriel River flowed even in summer all the way to the mouth of the canyon above Azusa. There was no other trout stream in southern California that could compare with it. The water was clean and clear, alternating between sparkling riffles and darker pools. Wherever the floods of winter had not scoured the boulders, alders and sycamores graced the banks, casting their shadows athwart the stream. The old wagon road wound its tortuous way in the canyon bottom, crossing and recrossing the rocks and river, and only seemed to make progress if it reached some bench or flat a little above the stream level and lined out across the sandy earth."

This picture of San Gabriel Canyon in the 1890's was drawn from memory by Stuart O'Melveny, son of Henry O'Melveny, one of the first to partake in the recreational delights of the canyon. In those days, mountain trout abounded in the sparkling waters of the river and its major tributaries. It was the lure of these speckled beauties that drew many an angler into the canyon.

As far back as the 1850's, when gold miners toiled in the East Fork and Eldoradoville was in its brief glory, the canyon of the San Gabriel was known for its fabulous fishing. A correspondent for the *LOS ANGELES STAR* (July 9, 1859) wrote: "This is really a beautiful cañon, and a more contented set of miners I have never seen . . . Should you think of visiting the mines at any time, I will promise you a mess of mountain trout which abound in the river."

During the last three decades of the 19th century, San Gabriel Canyon was known throughout Southern California for its abundance of mountain trout. The *STAR* of June 8, 1870, in describing an eight-day hunting and fishing trip into the canyon by a party of Los Angeles businessmen, stated, "They caught eight hundred fine trout, many of which measured from twelve to sixteen inches in length — the entire lot averaging

FOLLOWS CAMP, LARGEST RESORT IN THE SAN GABRIELS, 1896
Robinson Collection

nine inches." One avid Los Angeles fisherman was Henry W. O'Melveny, a prominent attorney and businessman, who became so enthralled with the trout-abundant canyon that, in 1889, he organized the Creel Club, dedicated to angling, gourmet eating and fellowship. Creel Club members, along with hundreds of others, enjoyed weekends and holidays fishing the waters of the lower East Fork as far up as Cattle Canyon.

It was inevitable, with the hordes of fishermen and vacationers in the canyon, that some enterprising entrepreneur would take advantage of the opportunity to cater to the canyon visitors. First to do so was William 'Lying Bill' Potter, who in the mid-1880's turned his Mountain View Ranch in the lower East Fork into a tourist resort. Vacationers were encouraged to camp at Mountain View for a small fee, and Potter opened a store to sell them fruits and vegetables grown at his canyon ranch. By the later 1880's, 'Potter's Ranch,' as it was generally known, was a thriving hostelry. The *AZUSA NEWS* (September 14, 1889) reported not only "swarms of people camped there who hail from all parts of the United States," but also that Potter's Ranch was becoming a favorite hangout of the 'busted lung brigade,' consumptives who had moved into the canyon to regain their health. After Lying Bill's death in 1903, his son Hayes Potter enlarged the resort and was the congenial host into the 1920's.

Potter's Mountain View Ranch was the first resort in the East Fork, but it was not the largest nor the most famous. That honor belonged to Follows Camp, founded by Englishman Ralph Follows in 1896.

The building of Follows Camp, located three miles up the East Fork on a broad bench south of the river, epitomized the miracle recovery story of Ralph Follows, who once had been given only a few months to live. Follows was critically ill with tuberculosis when his brother, Jack Follows, a sailor on the Liverpool to San Francisco run, learned of the supposed therapeutic powers of San Gabriel Canyon. Jack persuaded his brother to leave England and come to the canyon. Arriving in Azusa in 1891, the two brothers rented a cabin on the old Ferguson property in the lower East Fork. Ralph's health improved almost immediately, and it was not long until he was well enough to go into business. His little cabin became a short-order restaurant serving canyon travelers. In 1896 he purchased the Roberts property just upstream from the Ferguson place and began building a resort camp. The following year he married Jennie Heaton, daughter of canyon miner William Heaton.

Ralph and Jennie Follows had a knack for resort business, and it was not long until Follows Camp was the largest and best known hostelry in San Gabriel Canyon. "Good home cooking, a genial friendliness and welcome, good fishing and glorious scenery were the attractions." An early camp brochure boasted, "Camp life at Follows Camp is spent in real comfort. The tents are all good sized, well floored, cozily furnished and kept spotlessly clean. Where guests may prefer, homelike rooms may be reserved. The camp is provided with bath tubs, shower baths and plunge bath — and every modern arrangement has been incorporated to insure perfect sanitation and health throughout the camp. Pure mountain water is pumped by a private water plant which Mr. Follows has installed at enormous expense. Mountain fruits and vegetables, rich milk and fresh eggs are produced on the premises. The table is always supplied with fresh meats and seasonable delicacies, and ice is brought daily up on the stage."

To haul guests up the tortuous twelve-mile canyon road, Follows initiated the famous four-horse Follows Stage. Via this conveyance, thousands of vactioners from all over California made the trip in. The camp could accomodate as many as 200 guests at one time in its numerous cabins and tents. Quite a number of Western movies were shot around Follows Camp, with Ralph Follows and his four-horse stage, hired hands and even guests often appearing as 'extras.' Such early day movie stars as Mabel Normand and Fatty Arbuckle appeared in silent films with a Follows Camp setting.

Ralph Follows' major headache during the years of his camp's popularity involved the maintenance of the canyon road, washed out almost every winter. Mrs. Follows related in later years: "His life story was one of trying to keep that road open so he could pack people and things through, and out again. It meant appearing before the county board of supervisors and asking the Azusa merchants for help. Now it is possible to go in half an hour the distance it took a four-horse stage 3½ hours to travel. It was necessary then, though, to ford the stream some forty times, and even in the best of weather that was not an easy trip. In times of high water it was impossible." When the canyon road was impassable, the only way to Follows Camp was to take the old Woods Trail over the mountain, an all-day, narrow and steep passage via shank's mule.

In the later years of the camp, Follows operated a livery stable not far from the Santa Fe station in Azusa. Guests debarking from the train would be met at the station by the genial

Ralph Follows and transported the fifteen tortuous miles to the resort, via either the four-horse stage along the canyon road, or on horseback over the mountain on the Woods Trail.

Forest fires were another headache faced by Ralph Follows. The crackling-dry chaparral that covers most of the San Gabriel Mountains can erupt in a fiery holocaust that will spread with lightning quickness. A resort that takes years to build can be destroyed in a matter of minutes. Follows was very aware of this danger and was always ready to join the Forest Service on the fire line. He turned Follows Camp into a fire station in times of fire danger and taught his employees the best methods of fire-fighting. This preparation came to good use on September 12, 1919, when a blaze erupted near the East Fork Narrows, allegedly caused by a miner burning trash. The conflagration, fanned by 25 mile-per-hour winds and 90-degree temperature, quickly spread westward and for time threatened Follows Camp. Ralph Follows, his hired hands and even some of his camp guests were on the fireline, night and day, working to keep it out of the lower East Fork. Fortunately, a timely rain storm doused the flames before they reached the canyon resort, but 60,000 acres of the San Gabriel watershed were blackened.

The 1919 blaze was just a prelude to the terrible mountain holocaust of 1924, which began in San Gabriel Canyon and soon spread into the heart of the range. The fire first started at O'Melveny Flat in the main canyon on Sunday afternoon, Aug. 31, 1924. It was caused by a careless throwing of a match into the brush by picnickers. The conflagration spread with flashing quickness through the dry chaparral and timber. Its main thrust was westward, up West Fork and into the almost inaccessible heartland of the San Gabriels. One tongue of the fire, however, crept up East Fork, destroying most of Mountain View and reaching the very edge of Follows Camp. Back-breaking efforts by Ralph Follows and his fire crew barely succeeded in saving the resort. When it was over, the camp was intact, but the surrounding beauty of the East Fork was blackened, lessening the appeal to visitors for several years.

The fires clearly revealed the need for improved fire control techniques in Angeles National Forest. Forest Supervisor Rushton Charlton stated the main problem: "The greatest difficulty in fighting fires was the lack of roads over which to transport supplies to the men." As a result, the Forest Service in 1925 initiated a crash program to build fire roads and fuelbreaks. A road building program was initiated in the can-

CREEL CLUB, ca.1895
Robinson Collection

yon, which not only allowed quick access to the fire-threatened areas but made Follows Camp and other canyon resorts easily accessible to automobiles.

Ironically, the rebuilding and paving of the canyon road in 1925, which Ralph Follows had long urged, spelled the eventual downfall of Follows Camp. The first automobile had reached the resort in 1904, but the trip was a hair-raiser. No more. By the end of 1925 autos were a common sight at the camp. "The coming of the automobile made the trip to Follows Camp too easy, the romance was gone and soon the camp languished and died," lamented Sedley Peck. In 1926 Ralph Follows was killed in an automobile accident near Redlands. Jennie Follows kept the resort open for a few more years, but her toil was not worth the effort. People were not willing to stay overnight when they could drive in and out of the canyon so quickly. The end of Follows Camp as a tourist resort came with the Depression. The tents were removed, the cabins rented to canyon residents. Today, the once-famous resort, many of its buildings still intact and occupied by permanent residents, sits quietly above the tumultuous waters of the East Fork. However, Mrs. Peck still operates a modern restaurant and provides picnic areas for the day traveler.

There were other tourist's resorts on the East Fork, although none were as large nor famous as Ralph Follows' genial hostelry.

Nestled among the oaks and sycamores at the junction of the East Fork and Cattle Canyon was a beautiful little resort founded by Jay Gardner Scott in 1909. Scott originally called his resort Scott's Camp, but soon changed its name to Camp Bonita. A small lodge and store were built and tents erected alongside the river. Trout fishing at this elbow of the East Fork was unbelievably good, and it was not long until Camp Bonita was a favorite haunt of anglers. Scott was succeeded by Henry Willard, and under the latter Camp Bonita reached its zenith of popularity. Three times a week, and more often when traffic demanded, the Camp Bonita Stage would bring dozens of happy fishermen and campers to this verdant hostelry alongside the sparkling stream. Unfortunately, the camp's streamside location made it susceptible to the periodic flooding that often hits the canyon. The resort was damaged several times by high water during the 1920's and early '30's. The end came in March 1938, when most of the cabins and tents were washed away in the worst torrent in the recent history of the East Fork.

CAMP WILLIAMS, 1980
Walt Wheelock photo

A mile and a quarter downstream from Camp Bonita is Camp Williams, founded by Jim Williams in 1913. Williams became known throughout the canyon for his 'home brew' during the prohibition era, and activity that apparently escaped the eye of the usually vigilant 'feds.' Imbibers, along with the usual coterie of campers and fishermen, kept William's little tent camp well attended during the 1920's and into the '30's. Jim Williams, however, did not have his heart in his resort. He was more involved with his various mining claims, which included a tin mine up nearby Horse Canyon and several marginal gold prospects.

Camp Williams was acquired by Lyle and Ramona Johnson in the early 1950's. The Johnsons turned the old resort into a combination campground, trailer park, picnic area, store and restaurant. In the 1960's the Hunter family took it over, and they in turn were succeeded in 1976 by congenial Grace Melton who runs the resort today.

The old Ferguson property downstream from Follows Camp is now known as Shady Oaks or Follows Camp West. For many years it was a private retreat for the Ferguson family. After World War II it was briefly a religious retreat called 'Beulah Land.' In the early 1950's Tony Galleta bought it; gave it the name of Shady Oaks and built a campground, picnic area, store and snack bar. Today it is owner and operated by Joe Davison, who purchased it in 1979.

The record for the shortest life of any resort on the East Fork goes to Branscome Camp, once located alongside the river just east of Follows Camp. Branscome laid out his camp too close to the tumultous waters of the East Fork in 1913. He paid for his mistake the following year, when the little resort was totally washed away. Branscome remained in the canyon for a number of years but never rebuilt his camp.

Two small resorts in Coldwater Canyon, a tributary of the East Fork, need to be mentioned.

One of the most beautifully-situated hostelries in all the mountains was Weber's Camp, founded by J. P. Weber in 1906. Deep in Coldwater Canyon, shaded by tall Jeffrey pines, live oaks and alders, Weber's Camp was a delight to its small but loyal clientele, many of whom returned year after year. The resort consisted of a main lodge and a half dozen housekeeping cabins alongside the icy-cold stream. Trout fishing and deer hunting were featured, as well as hiking. Weber himself built a steep trail from his camp to the summit of Old Baldy. Parties went up one day, stayed overnight at William B. Dewey's

'Summit Inn,' and returned to Weber's Camp the next day. Rates at the camp were $2 per day for room and board, or $10 per week. Or, a person could rent a tent for $4 per week. The resort reached its zenith of popularity in the early 1920's — its demise came with the depression.

A half mile upstream from Weber's Camp, Jake and Flora Widman homesteaded a scenic plot and built a log cabin in 1913. Later they erected more small cabins and took in paying guests. Not really a commercial resort, Widman's was never-the-less a favorite retreat for those who relished a wilderness setting.

Building Highways

"Store your auto at our Azusa garage," read a Coldbrook advertisement in 1911, "and the stage will carry you to the camp." Indeed, the average motorist risked his automobile and sometimes his life to negotiate the tortuous San Gabriel road. The route followed the canyon bottom and was unpaved, very rough in places. It forded the river without benefit of bridge nineteen times between Azusa and the Forks. "As the year's flood season would render it useless and after spring freshlets had subsided, the work had all to be done over," recalled Cornelius Smith, canyon historian and secretary of the Azusa Chamber of Commerce. As mentioned earlier, Ralph Follows took the initiative every year in persuading Azusa merchants and other canyon resort owners to repair the road. "A yearly sum of between two and three thousand dollars was raised and expended on the canyon road without a single penny to our aid from outside sources," related Smith.

It was obvious that any major highway construction in the San Gabriel Canyon would require financial support from outside the local area. The great mountain fires of 1919 and 1924 finally provided the motivation, and work commenced on a paved highway up the canyon in 1925, financed jointly by the U. S. Forest Service, Los Angeles County and the State of California. By year's end the new highway had reached the Forks, and a preliminary bridge spanned the river to give access to the East Fork resorts.

Unfortunately, the new canyon thoroughfare was built too close to the river and, like the old road, suffered frequent washouts. What was needed was a high, washfree route, but the cost for such a canyon-side highway was considered prohibitive.

But fortunately, necessity became the mother of creation. In the late 1920's, the Los Angeles County Flood Control District and the City of Pasadena both planned to build major dams in San Gabriel Canyon. As these concrete bulwarks would block the canyon-bottom road, a new 'high-line' highway well

ROAD REBUILDING AFTER 1980 FLOOD WASHOUT
Walt Wheelock photo

up on the west slope would be required.

In 1926 and 1927 a spur railroad line was built slightly above the canyon bottom to the site of the proposed dam at the Forks. It even had a tunnel bored through a ridge, just below the present site of the Morris Dam. As this never carried much traffic, it also became the major entrance way to the Forks, and it was sometimes possible to even sneak through the tunnel with car and cut off a half mile of curving road. The dam at the Forks was found to be on an active earthquake fault and so that dam was never built. The railway route remained the main access route, until it was blocked by the construction work on the Morris Dam of the City of Pasadena.

In the meantime, the proposed 'high line' thoroughfare was surveyed by county engineers in 1930 and construction got underway in 1932, financed by four agencies: the Los Angeles County Flood Control District, the City of Pasadena, the Forest Service and the State of California. The new route, snaking along the west canyon wall, required an immense amount of excavation, as much of the rock proved to be decomposed granite, the slope was steep, and recurring slides necessitated frequent repairs. The canyon was a beehive of activity, as the two dams — Pasadena's Morris Dam, and the County's huge earth fill dam, that replaced the earlier planned concrete dam on the earthquake fault — rose from the riverbed while the mountainside highway inched its way northward, The preliminary roadway was blasted and graded to the Forks by the end of 1933, although it was not paved and ready for public use until 1935. Widened and improved in recent years, it is still the only thoroughfare up the main canyon.

While the high line route was under construction in the main canyon, not one but two major highways were being built into the East Fork. The Glendora Mountain Road, climbing the divide between San Gabriel and Little Dalton canyons and then dropping into the East Fork near Camp Bonita, was completed in August, 1931. Far more spectacular was the County's proposed East Fork Highway, to follow the East Fork all the way to its head, then climb over Blue Ridge and drop to Wrightwood on the north slope of the range. This would be, in the word of one county supervisor, "The most scenic and breathtaking mountain highway in California."

Construction of the East Fork Highway commenced in 1929. County prison crews did the lions share of the work, beginning near Camp Bonita and working west down the canyon toward the Forks, then blasting and clearing their way from the

Cattle Canyon junction northward towards the Narrows. Here, where the walls of the East Fork close in and form the deepest mountain gorge in Southern California, a concrete bridge was built and a tunnel chiseled out of sheer rock. But this was as far as the projected cross-range highway got. On the night of March 1-2, 1938, an unprecedented rainstorm hit the mountains, causing a savage torrent to boil down the canyon. Literally overnight, the East Fork Highway was cut to pieces. Damage was so very extensive that the new highway was abandoned, its remains standing as an object lesson for the foolish engineers who planned a highway too close to the fickle river. Since 1938, 'The Bridge to Nowhere,' just below the Narrows and visited only by hikers, has stood as a mute reminder of the power of mountain water.

The idea of a road up the East Fork and across the mountains was revived after World War II. Its purpose would be more than just scenery and recreation; the Cold War with Russia was on and the highway was planned as an escape route for the millions of people who would be "trapped" in the Los Angeles Basin in event the cold war turned hot. In 1954 the County Road Department, again utilizing prison labor, began construction of this new highway up the most spectacular chasm in the San Gabriels. This time, to avoid the flood hazards of the canyon, the roadway was etched high along the west wall. Original plans called for completion to the Blue Ridge and a junction with the Angeles Crest Highway by 1977. Working daily shifts, the prisoners from County Detention Camp 14 had four and a half miles of superb highway carved out of the mountainside and graded by 1969. Then the County ran into opposition from the budget-cutters, who felt that the mountain road was too expensive, and from conservationists who wanted to preserve the upper East Fork in its wilderness state. Work abruptly stopped, and not a shovelful of dirt has been turned on the mountainside highway during the past decade. Barring a complete turn of events or some unforeseen emergency, the road will never be completed. The excuvation scars remain an ugly stain high on the western slope of the canyon, a defacing monument to plans that went astray.

Sheep Mountain Wilderness

Beyond the road's end at Heaton Flat, the East Fork changes character. Sidewalls steepen, the gorge narrows, the river becomes more turbulent and you enter the most impressive chasm in Southern California. The scenery here is monumental, on a scale seen nowhere else in the San Gabriels. The rise from the floor of the Narrows (2800') to the top of Iron Mountain (8009') is 5200 feet in one and three-fourth miles! This is nature in its grandest proportions — at least by Southern California standards.

This upper East Fork country and its surrounding peaks — 9399-foot Mount Baden-Powell on one side, 10,059-foot Old Baldy on the other — is the summer home of one of the largest herds of Nelson Bighorn Sheep in the state. To protect these "statuesque masters of the lofty crags," and to preserve some of the wildest mountain country left in Southern California, the federally-designate Sheep Mountain Wilderness is being established. The new wilderness area, which needs only a final approval by Congress to become a reality, will include the entire headwater region of the East Fork and its adjacent mountain country, some 45,000 acres in all, to be forever preserved in a natural state.

The Sheep Mountain Wilderness has been a long time in reaching fulfilment. The first proposal to preserve the East Fork headwaters was made by the Azusa Chamber of Commerce in 1935. The 'San Gabriel Gorge Primitive Area' proposal was supported by the Sierra Club and several hiking organizations, but it never got beyond the talking stage.

The idea lay dormant for some thirty years, until the Sierra Club renewed its support of the wilderness proposal in the late 1960's. A group of concerned conservation-minded citizens formed the Sheep Mountain Task Force in 1972, and began working energetically to get a wilderness designation for the area. Their efforts were rewarded in October 1976, when President Gerald Ford signed a bill creating the Sheep Mountain Wilderness Study Area. This was followed by three years of scientific study, public hearings and voluminous cor-

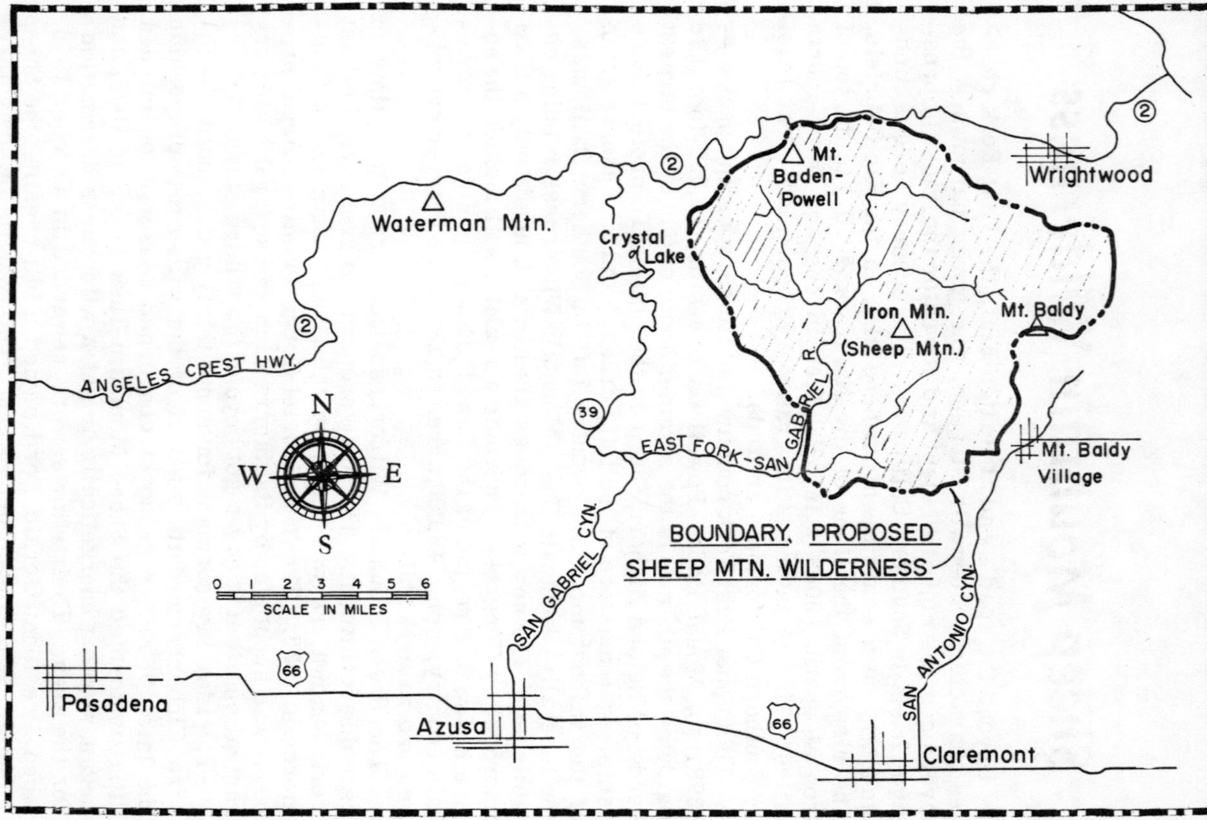
Wrightwood
Mt. Baden-Powell
Waterman Mtn.
Crystal Lake
Iron Mtn. (Sheep Mtn.)
Mt. Baldy
Mt. Baldy Village
ANGELES CREST HWY.
EAST FORK-SAN GABRIEL R.
SAN GABRIEL CYN.
SAN ANTONIO CYN.
BOUNDARY, PROPOSED
SHEEP MTN. WILDERNESS
N
S
E
W
0 1 2 3 4 5 6
SCALE IN MILES
2
39
66
Pasadena
Azusa
Claremont

respondence. The major dispute among wilderness supporters and opponents was over the size of the proposed wilderness. Conservationists asked for 52,000 acres, some hunters and ORV enthusiasts wanted less than 20,000 acres set aside. The Forest Service proposed, a 30,400 acre conclave surrounding the East Fork headwaters.

While the Sheep Mountain Wilderness seems assured, its size is not, as of this writing, been determined. In November 1979, Congressman Jim Lloyd of the 35th district, encompassing the Claremont-Pomona area introduced HR 5541, the Sheep Mountain Wilderness Bill, asking for 45,000 acres. Lloyd's bill would establish a wilderness area stretching from Mount Baldy west to Mount Baden-Powell, and include all of the East Fork headwaters and upper tributaries. It remains to be seen as to whether Congress will accept HR 5541, or opt to support the Forest Service's more modest proposal.

MODERN-DAY PLACERING ON THE EAST FORK
Walt Wheelock photo

The East Fork Today

Old-timers would hardly recognize the East Fork of today. The magnificent stands of live oak and alder that once lined the river are mostly gone, replaced by a broad swarth of naked boulders, a legacy of recent floods. The mining camps have disappeared; the old resorts are only shades of their former selves, catering to a different, more mobile clientele. The historic East Fork wagon road that for so long followed the river, crossing and recrossing it innumerable times, has been replaced by an asphalt highway well up on the canyon slopes.

Yet, some things have not changed. The East Fork remains a beautiful mountain canyon with a sparkling stream, green-covered hillsides, and Old Baldy's gray crown (seen from the lower canyon) an imposing backdrop. As in decades past, hundreds enjoy weekends and holidays in the canyon, camping, picnicking, hiking, fishing and panning for gold. The East Fork has long been a favorite recreational area for the multitudes who live in the San Gabriel Valley and other parts of Southern California, and it remains so today.

The visitor to the East Fork is literally steeped in history. Probably no other area in the San Gabriel Mountains is so richly endowed with the heritage of yesteryear. To guide the canyon visitor and to point out the historic haunts and scenic attractions of the canyon, the following mileage log is offered, beginning at the junction of the main San Gabriel Canyon Road (Highway 39) and the East Fork Highway.

ROAD LOG

MILEAGE partial	MILEAGE total	DESCRIPTION
0.0	0.0	You cross the East Fork Bridge, a modern flood-proof span of steel and concrete, completed in 1949. High ahead, above the chaparral-clad canyon slopes, looms Old Baldy (10,059').

CAMP FOLLOWS STAGE, RALPH FOLLOWS STANDING
Robinson Collection

0.7 0.7 Burro Canyon crossing. An old pack trail once run up canyon and over the ridge to the mines of the upper East Fork and Iron Fork. There was considerable mining in this tributary canyon, and you can still find tunnel remains and remnents of mining tools with diligent searching. John MacCaslin, miner and packer, had a cabin at the canyon mouth, now buried under tons of rock of the road fill.

1.3 2.0 Camp Oak Grove turnoff. A bridge crosses the river to the broad, oak-shaded bench that was once known as Mountain View, the home of the Arviso family, earliest known residents of the East Fork, and later 'Lying Bill' Potter's Ranch, first resort in the canyon. The camp belongs to Los Angeles County now, and is the home of Julius M Klein Conference Center, a human relations retreat.

0.1 2.1 Susanna Canyon crossing. Named for Henry C. Roberts' wife. Miner Ed Fricke lived at the canyon mouth for other forty years and had several gold prospects up the canyon.

0.5 2.6 Graveyard Canyon crossing. Named for ancient Indian burial mounds two miles up the canyon. Considerable early-day gold ming was done here. An old pack trail once climbed the canyon and dropped to the Iron Fork mines.

0.1 2.7 Shady Oaks, or Follows Camp West, turnoff. A bridge crosses the river to this restaurant picnic area, once the William Ferguson property. With a keen eye, you can still make out traces of the Ferguson hydraulic mining effort on the slopes above the camp.

0.3 3.0 Follows Camp turnoff. Sideroad fords the river to a restaraunt and picnic area owned by Mrs.Sedley Peck. Follows Camp, founded by Ralph Follows in 1896, was the largest and most famous resort on the East Fork. Before that it belonged to Henry C Roberts, and the store he built in 1861 still stands, the oldest structure not only in the canyon, but in the entire San Gabriel Mountains. Just up-canyon, to the left of the main highway, can be seen the scars of Roberts' hydraulic mine.

0.3 3.3 High Road turnoff, to your left. This broad well-graded highway, constructed by prison labor from County Detention Camp 14, was projected to cross the mountains to Wrightwood. It never made it. The road abruptly ends high on the west canyon wall, 4½ miles from start.

0.2 3.5 Detention Camp 14, to your left. County prisoners here did the lion's share of road construction and maintenance in the canyon from the 1940's until last year, when the camp fell victim to Proposition 13. Today, the camp is run by the California Conservation Corps, a state reclamation agency.

0.2 3.7 Cape Horn Bridge. You cross to the south side of the canyon. Back to your right is the narrow, heavily-wooded gash of Cape Horn Canyon. A quarter-mile up the canyon is Cape Horn Falls, once a favorite hike from Follows Camp.

0.2 3.9 Camp Williams. This thriving settlement boasts restaurant, store, trailer park, and camp and picnicking areas. Jim Williams founded the camp in 1913; today the proprietor is genial Grace Melton.

0.8 4.7 Oaks Picnic Grounds, to your right in a pine and cedar-shaded cove, with stoves and picnic tables. It should be called Troubadour Camp, for just above was the cliff-cabin of Troubadour Rodriguez, a Yaqui Indian miner about whom canyon legend ran riot. Besides recovering gold from several hidden prospects, Toubadour tended cows and cultivated an alfalfa field across the canyon, and hunted game with a sling shot.

0.4 5.1 Glendora Mountain Road turnoff, to your right. This highway completed in 1931, climbs the south ridge and divides, the left fork going to Camp Baldy, the right down to Glendora. The road offers fine panoramic views over the East Fork country.

0.2 5.3 Cattle Canyon Bridge. To your right a few hundred feet up Cattle Canyon, is site of Camp Bonita, a popular fishing resort in early years of this century. Left, on the north bank of the river is Eldoradoville Picnic Grounds, the most storied spot on the East Fork. Here once was the raucus mining town of Eldoradoville, washed away lock, stock and barrel in the great flood of 1862. The concrete foundations here are what's left of Camp Bonita Ranger Station, obliterated in the 1938 flood.

0.6 5.9 East Fork Fire Station, up to your right. The Forest Service wisely built this station well above the river, away from the threat of floods.

0.1 6.0 Road's End. Beyond the locked gate is the domain of the hiker, the area of the proposed Sheep Mountain Wilderness. On foot, you may follow the old East Fork Road to Heaton Flat, the site of Billy Heaton's cabin, orchard and gold mine, a half mile. Five miles farther, after several foot-wetting river crossings (dangerous in time of highwater), you reach the 'Bridge to Nowhere' and the spectacular Narrows, a steep-walled gorge more befitting the High Sierra than the San Gabriels!

'BRIDGE TO NOWHERE', EAST FORK NARROWS
John Robinson photo

References

This book is based on factual references. Written for the general public rather than as a scholarly dissertation, it is not footnoted. Those wishing documentation should consult this writer's "History of San Gabriel Canyon," in five parts, which appearedin THE POMONA VALLEY HISTORIAN, Spring 1977, Summer, 1977, Fall 1977, and THE SAN ANTONIO HISTORIAN (same journal, name changed), Winter 1978 and Spring 1978.

The following sorces havebeen the most valuable in learning of the saga of the East Fork.

Books

Leadabrand, Russ, GUIDEBOOK TO THE SAN GABRIEL MOUNTAINS (Los Angeles, 1964).

Robinson, John W., MINES OF THE SAN GABRIELS (Glendale, 1973).

Robinson, W. W., THE FOREST AND THE PEOPLE (Los Angeles, 1946).

Saunders, Charles Francis, THE SOUTHERN SIERRA OF CALIFORNIA (Boston, 1923).

Show, S. B., HISTORY OF ANGELES NATIONAL FOREST, U. S. Forest Service typescript (unpublished, 1945).

Wheelock, Walt, SOUTHERN CALIFORNIA PEAKS (Glendale, 1973).

Periodicals

Gay, Thomas E. and Samuel R. Hoffman, "Mines and Minerals of Los Angeles County, California," CALIFORNIA JOURNAL OF MINES AND GEOLOGY, July-October 1954.

Guinn, J. M., "The Gold Placers of Los Angeles," LAND OF SUNSHINE, July 1896.

Hoffman, Abraham, "Mountain Resorts and Trail Camps in Southern California's Great Hiking Era, 1884-1928," SOUTHERN CALIFORNIA QUARTERLY, Fall 1976.

Peck, Sedley, "Colorful Old Days on the Upper San Gabriel," TRAILS MAGAZINE, summer 1938.

Robinson, John W., "Eldoradoville: Forgotten Southern California Mining Camp," POMONA VALLEY HISTORIAN, Spring 1976.

Robinson, John W., "The Great Hydraulic Race," DESERT MAGAZINE, October 1977.

Rolfe, Frank, "Gold Mining Days in the San Gabriel Canyon," SOUTHERN CALIFORNIA QUARTERLY, June 1953.

Vernon, Charles C., "A history of the San Gabriel Mountains," SOUTHERN CALIFORNIA QUARTERLY, March, June, September and December 1956.

NEWSPAPERS

Azusa HERALD, October 20, 1937 (special issue on the San Gabriel Canyon.

Azusa POMOTROPIC, 1889-1928

Los Angeles STAR, 1864-1868.

Los Angeles SEMI-WEEKLY NEWS, 1864-1868.

SOUTHERN VINEYARD, 1858-1859.

MANUSCRIPT COLLECTIONS

Will Thrall papers and photographs, Henry E. Huntington Library, San Marino.

Sedley Peck papers and photographs, Follows Camp, East Fork of the San Gabriel Canyon

Index

FLO FLO PECK AT HER RESTAURANT, CAMP FOLLOWS
Walt Wheelock photo